CONSCIOUSNESS PROBED

THOUGHT EXPERIMENTS AND SPECULATIONS ON CONSCIOUSNESS, AND IMPLICATIONS OF CONSCIOUSNESS IN MACHINES

SUMEET S. NAVALKAR

ISBN 979-888606242-7

Contents

Preface

An Analysis on I.

What is consciousness? Consciousness is a word used to describe awareness- awareness of the self, awareness of the world.

Are all living beings conscious? It is difficult to say. We will consider plants. Plants are aware of their surroundings. They are because a plant's roots know they have to penetrate the soil. Likewise the shoot knows that it has to grow towards the sun. Plants are a complex organism. They might have consciousness. What about bacteria? A bacterium too is aware of the surroundings. The bacterium takes in specific nutrients from its surroundings for its growth and reproduction. When the environment or the surroundings become unfavourable for a bacterium's growth, some species form spores that form a protective covering around the cells of the bacteria. In the spore, the bacterium is in a dormant stage meaning when the conditions become favourable again, the spore coat disappears and the bacterium resumes its routine with normal life. The bacteria, like the plants, can thus be said to be aware of their surroundings. What about the animal kingdom then? What about cats, dogs, cows, pigs and many other animals that we see in our daily life? The consciousness or awareness of animals to their environment is more obvious. The animal kingdom, in general, reacts faster to its environment than the plant kingdom. The movement of the animals in reaction to external stimulus is what tells us that animals too are conscious of their surroundings. However, when we mean consciousness, it is not only awareness about the surroundings; it also means awareness about one's own self.

In science, there is a way to determine whether an animal is self-aware. It is the mirror test. An animal is faced with a mirror. Almost all animals feel that their own reflection, which they see in the mirror, is another animal. They do not pass the mirror test. Chimpanzees on the other hand, though too initially perceive their mirror image as another chimpanzee, in course of time, realise that it is their own self that they are looking at. They pass the mirror test, and hence they are termed as self-aware. There are few other animals which pass the mirror test. Scientists suggest that though the mirror test is a proof of self-awareness, it is basically a measure of how self-aware an animal is. Animals who fail the mirror test cannot be discarded as being non-self-aware.

We shall consider this. A monkey does not recognise itself in the mirror. It may react aggressively towards its reflection feeling that the reflection is another monkey which can be a threat. This behaviour itself demonstrates the monkey's self-awareness. How? If the monkey is not self-conscious, why would it recognise another monkey as separate and a threat? I feel recognition of something or someone as a non-self in itself forms the root of self-awareness. Extending this logic, other animals too, which fail the mirror test, can be said to be self-aware. Plants and bacteria too, in this respect, are self-aware. This, however, is debatable. Someone might point out that absorbing nutrients from the surroundings, as is done by plants and some bacteria, is not the same as recognising the surroundings as non-self. Plants and bacteria fall under this category. The absorption of nutrients and water from the surroundings is a plain mechanical process. The plants and bacteria do not have a choice, we can say. For example, water transpires or evaporates from the leaves creating a deficit which is replenished by absorption of water from the roots. The roots absorb water by the mechanical process of osmosis. If so, the reactions that animals show towards their environment also peel off to plain mechanical processes. Some physicists go to the extent of saying that free-will is an illusion. That is, every reaction of any human being is predictable, and that too, precisely (if of course we have a computer that is extremely fast and powerful to do the calculations and the prediction; this computer may however not be a reality in far future too). Depending upon the strength of the stimulus, the reaction it causes in our nerve cells can be calculated. Taking into account other factors, a reaction of a particular human being to a particular stimulus can be precisely calculated and thus predicted. This brings our own behaviour and reactions to the environmental stimuli down to just simple mechanical processes. So if we with our mechanical reactions are still self-aware, we can say the same thing about plants or lower organisms. Once an organism becomes aware of its non-self surroundings, it can be said to become self-aware.

There is one more argument that we can consider. Higher animals with their brain and nervous system (as opposed to plants and bacteria) are able to think and feel themselves physically. They thus know that what they cannot feel physically when touched is not a part of their body. An organism with a nervous system is aware of the extreme ends of its body. But can this be said of plants? Yes, it can be. An acasia tree when it is under attack (eaten by a giraffe), produces a chemical which makes itself inedible. It is a self-defence mechanism. The acasia realises that it is being eaten in spite of not having a nervous system. Thus we can infer that nervous system is not a necessity for an organism to be aware of its body though it can still be argued that the reaction of the acasia is a mechanical process. So we consider one more example. Some plants depend upon the animal kingdom for their pollination. Plants develop nectar, fragrances and attractive colours in their flowers with the sole intention of attracting insects or animals which carry out pollination for the plant (and many times propagation of seeds through their sweet fruits). This is an example of plants being aware of their surroundings- surroundings which are living and yet not a part of their body. Isn't this self-awareness which stems from awareness about non-self? Extending this logic to all organisms, can't we conclude that every living organism is aware of itself directly or indirectly? It might be safe to say that even in the absence of a nervous system, an organism is conscious of itself; with the presence of a nervous system, this awareness becomes more obvious to us. This awareness is what consciousness is. Each organism can thus be said to have consciousness.

We move ahead now to the computers. Complex and learning computers have already been built. Can a computer or a robot ever become self-aware? Computers (or robots) which learn from their environment are a reality. An intelligent robot which learns from the environment and reprograms itself can finally realise the capability and limitations of its own self. Though it does not possess a nervous system, it can become aware of its bodily dimensions (in other words- its body limitations in space). Once the computer or robot knows which part of the environment belongs to itself and which doesn't through trial and error learning methods, it will become aware of the self and non-self environment. A computer realising this then becomes self-aware and thus conscious.

In the following text, we will speculate on such consciousness arising from a 'non-living' entity along with consciousness in general. We will also speculate on consciousness from different perspectives, and in the process contemplate the implications of robotic consciousness.

It is important to note that we can contradict what we have already said here; we may even contradict our own contradiction. It will be a discussion, and I will let it move in any direction as we penetrate deeper into the subject. Though the preceding text implied that bacteria and plants have consciousness, in the text that follows, we will consider only human-like consciousness- that consciousness which comes with the process of thinking. Under these considerations and in the absence of a nervous system (or something that gives the organisms the ability to think) in plants and bacteria, we will have to consider these organisms as non-consciousness.

Note:

In the following text, you will encounter two kinds of I's. One is I; the other is 'I'.

I has its usual meaning. 'I' stands for the self in an organism.

CHAPTER I

I happened to come across a hypothesis or a theory that states that a human being will be able to live forever as a machine (computer) if all his memories (inclusive of experiences) are downloaded on a memory disc. To clarify, it will not just be a set of memories. It will be a computer that will also be able to think like a human. It will be a fast and powerful machine to be able to think, and it is supposed to think like that particular individual depending upon his experiences. Knowing this, I had my doubts- not on the existence of such a machine in future, but whether this computer will be able to make an individual immortal.

Suppose, all my thoughts, memories and experiences, and whatever it takes, are downloaded on such an advanced computer (capable of thinking, speaking, walking etc.) at this moment. Does it mean that now I am living as two individuals at the same time? Or is it that the computer and the human are the same individuals? Will I start feeling that I am a computer? Will the computer start feeling that it is a human? The second option might become a possibility. But still, I think, I will not be able to relate to the computer as 'I'. 'I' is not just my thoughts, experiences and memories. 'I' is an abstract something that makes me up. (I can say that 'I' is contained in the limits of my body. However, what is MY BODY? How can we differentiate that from another body? The nervous system with its sensory input does that for us.) We will consider something else.

Now suppose I die. And instantly after my death, all my memories are downloaded on an advanced super computer. Now this scenario is different than the above one. In the previous case, I won't be able to relate to the computer as me. We are different. In the current case, I am dead and the essentials that made up my 'I' are downloaded on a computer. Here, there is only one 'I' left. Originally, it was a human; now it is a machine. For all practical purposes, for all my friends and relatives, I will still be alive- thinking, walking, talking, loving- just like when I was a human. By 'loving', I mean only the programmed DISPLAY of the human emotions. The machine will behave (at least externally) like a human. If it can think, it will think that 'I' am still alive. It will be self aware like me. So what about consciousness? We can say that where there is 'I', there is consciousness. Now the thing is that the machine thinks that 'I' am still alive. But does that mean I am still alive? Maybe- because though I remember I was once dead, death would then have become a reversible process.

Then what is the real difference between the first scenario where I am alive and the second scenario where I am dead. In the first case, as I am alive, the existence of someone like me (or another me) suddenly reiterates the need for me to be I, and as there are two individuals (one human and one machine), the difference is apparent. The human knows he is human. The machine, though it thinks it is human, is intelligent to know that it is not. This may change the psychology of the machine. In the second case, I am dead, and hence though the machine knows that it is a machine, there is no me to give it a complex (superiority or inferiority) or competition. In this case there is no pressure for the human or the machine to be more human. In the first scenario, the memories of the human and the machine begin to diverge the moment the machine is formed; in the second case, there is only one set of memories that grows as time flows; initially, that set belonged to a human, now it belongs to a machine. So when two sets of memories (though almost exactly same) exist at the same time, there are two different 'I's. So does it mean that when there are two different brains (or brain-equivalents), there are two different 'I's? The answer seems yes. If I am operated on and I am attached with 2 more hands and 2 more legs, will there be two 'I's? No, right? So the brain is the seat of 'I' for the purpose of this discussion. And if there is consciousness where there is 'I', then the computer (with a CPU, but without a brain) can also become conscious. In both scenarios (where I am alive and where I am dead), the machine is conscious and it thinks that it is 'I' and it is me. But in the first case, I am there, and so logically the machine is not me. In the second case where I am dead, there is only one consciousness that exists in the form of a machine and since there is no way to compare two different 'I's, we can safely say that though I am dead, I am still alive in the form of a machine.

Some of what I have written above can be debated. So let's consider.

We have created consciousness in the form of a machine in both cases whether I am dead or I am alive. If when I am dead, the machine can be assumed to be a continuance of my own self, then why not in the first case? After the machine is created, on what basis can anyone say that it a lesser 'I' than the human 'I'. Just because one consciousness resides in a tin and the other in flesh, can it make one consciousness inferior to the other, and if not, who is the real 'I'? Suppose, after the creation of the machine, I start behaving schizophrenic? The machine will not have schizophrenia. Schizophrenia can be genetic and hence a result of proteins. Since the machine does not produce proteins, it will not become schizophrenic. (It can be programmed to act schizophrenic, but that is totally different.) Also in the machine, there will be nothing equivalent of brain damage. So we come back to the point where I start acting schizophrenic. Then what? The human consciousness is now damaged since my brain (and perhaps body) has two or more different 'I's as a result of schizophrenia's altars (altars are different personalities that the patient having schizophrenia exhibits). The machine has the same original 'I'. Now since the human has become schizophrenic, doesn't that give the machine's 'I' more credibility? Doesn't that mean the machine is now more 'I'-like than the human 'I'? Then if I hadn't had schizophrenia, how does that make the machine 'I' less credible? And if it doesn't, then who is the real 'I'- the human or the machine? By creating the same consciousness in a machine, we have given rise to a parallel self. It is like having two realities in the same universe- a parallel universe within a single universe. If we still need to differentiate the two, we can say that one consciousness is the copy of the other. But we cannot say one is real and the other is not. One is as real and as credible and as authentic as the other. Just because one came before the other does not make it any more real.

(Now, what the hell is consciousness? Does consciousness only mean memories, thoughts, experiences and capability to think? Or is it just 'I'? And if it is 'I', does it just mean self-awareness? We will review what we already have earlier, but please read on. If 'I' just means self-awareness, then perhaps only a few animals like man and chimpanzees who recognise oneself in the mirror are self-aware. Rest of the animals like some lower monkeys- when they look into the mirror- happen to think that the reflection is another monkey. Chimpanzees, though initially think the same, after a while, realise that it is a reflection and not another chimpanzee. This self-recognition makes us self-aware, and if this is what is consciousness, then except a few, it makes all the animals on this planet not conscious. Consciousness then becomes the quality of the elite species. A rat is not conscious. A dog is not conscious. Now if we consider that 'I' is not only about self-awareness, then that brings a lot of species under the umbrella of consciousness. Any animal tries to escape from pain. The nerve signals tell the brain that its body is in danger. Any individual of any animal species that tries to preserve its own life can be assumed to be conscious. If so, the Acacia tree that produces toxin when it realises that it is being eaten can also be said to be conscious. Now, let's consider the first alternative that of consciousness meaning memories, thoughts, experiences and capability to think. If this is what comprises consciousness, then consciousness does not pervade through human babies. Moreover, human babies are also incapable of passing the mirror-test. For a patient of Alzheimer's disease where memories can fade to the point where the person can forget his own name, or due to brain degeneration can lose the ability to think, does consciousness leave such a human being before his death? There is no 'I' in such people. So if we consider this definition of consciousness, it seems that consciousness comes with a thinking brain, and if consciousness does not come free with life, then it belongs only to some higher animals including man and then too it does not pervade all individuals in the elite species. Which definition of consciousness is correct; which is wrong? What if after the death of an Alzheimer's patient, all what remains of his brain is downloaded on a computer and he is made to live forever? The machine perhaps will not be conscious of its own existence. But what does it imply? Can life go on without consciousness? If memories, thoughts, experiences and capability to think together form consciousness, we can try asking a few questions to an extremely sleepy person, and can conclude that consciousness has left him. Does consciousness leave us every night? We digressed a bit and tried to understand what consciousness means. But the attempt left us with more questions. The word, 'consciousness', will be used in the text that follows, and we will assume its meaning to what we meant before this paragraph. Please do not forget that this whole paragraph is in parentheses. We will continue from where we left before the bracket opened.)

Comparing two almost identical 'I's reminds me of twins. Monozygotic or identical twins are naturally occurring clones. Well, they are not exactly clones. Even though the chromosomal DNA is exactly same in both individuals (2 if it is twins or 3 if it is triplets, 4 if it is quadruplets and so on), some other cellular DNA differs in these individuals. But even when they are so similar, their 'I's are different. Even if they live in the same circumstances- same home, same school etc.- there is bound to be some difference. Their memories, their thought processes differ because they cannot be at the same exact location all the time. Even if they are, one twin might be to the left of the other, for example. These minor differences also lead to discernible changes in the personalities of identical twins. From this, we can infer that consciousness or 'I' is not just genetic in nature. Not just. The 'I' to a large extent also depends upon genetics. It can be observed when identical twins are compared to non-identical (also known as fraternal of dizygotic) twins. This is not about physical appearance of fraternal twins. The difference in behaviour between fraternal twins is greater when compared to difference of behaviour between identical twins. So both, genetics and environment, play a part in constituting the 'I' of any individual.

On the basis of the above, we can say that consciousness is both genetic and acquired. So if we apply this to one of the previous scenarios, what does it imply? The computer which is the exact copy a human being has no genes- no genetics is involved. All of its 'I' is directly acquired. If consciousness is product of genetics and environment, then can the machine have consciousness? Now another complication puts its head up. If consciousness is the product of genetics and environment, then all life on earth has consciousness. In fact, a bacterium can then be said to have an 'I'. And this logic can be stretched to apply to every cell in the human body. Every cell in the human body can have a separate 'I'. Now this becomes utterly ridiculous. Someone can argue that this logic is OK and that the sum-total of each 'I' in our body makes up the single macro 'I' that we understand or know. But is it? If my brain is transplanted in another human's body, obviously my 'I' will be transferred to that body. Those micro little 'I's from the recipient's body will be crushed by the brain transplant. If so, those micro little 'I's did not exist in the first place. So we must say that though 'I' can depend upon genetics and environment, 'I' or consciousness can still exist without genetics. This means a machine can have consciousness without the necessity to possess genes. This means consciousness can exist without life (life as we know it). But can we also safely say that life can exist without consciousness? Well, the answer to this can be 'yes'; bacteria or lower organisms cannot really be said to possess an 'I' (though we have seen earlier that they can possess some sort of consciousness). As already stated, consciousness can exist without life (if it can exist within a computer), and if so, perhaps consciousness needs only information to exist. And there is so much information in the universe (our telescopes can capture radiation from the time of the big bang- these are all memories of the universe). Can't that information lead to the existence of consciousness without the need of a physical body? This is something to think about. Well, on second thoughts, perhaps our definition of life itself is wrong. Maybe the computer could be said to have life since it is self-aware though it does not contain anything physical that we can attribute to as living. On the other hand, a bacterium, which lives just by chemical interactions with its surroundings without any knowledge of itself or life, can be considered to be non-living though it has the 'genetics'. This argument about what is life and what is not can become too complex and never-ending. So let's go back in search of 'I'. Human consciousness resides in the brain. Computer consciousness resides in a chip. Where does the consciousness of the universe reside? Perhaps, the universe as a whole acts as the brain. But does it have only memories? For consciousness to exist, the universe should also be able to think. But what is thinking? For a human, it is information transfer between neurons based primarily on memories or modified memories- basically intermingling of memories (memories do not necessarily mean something that has already happened- whatever the brain conjures up becomes a memory). For a computer chip, it is circuits instead of neurons. For the universe? Perhaps, for the universe, it is much simpler. The memories are the causes which give rise to thoughts which are the effects.

Just some time back, we saw the scenario of twins. Let's consider clones now. The technology of cloning at present involves the necessity of a donor (the person whose clone is to be created) and an egg (from a female) and a womb where the foetus (baby- in layman's terms) can grow. If my clone is to be created now, one of my body cells, an egg from a female and a womb will be required. After nine months, my clone will be born, and he will only be an infant. However, for the discussion here, we will not consider such a cloning procedure. We will consider a futuristic cloning technique where a clone created will be my present copy. He will be my age, my height, my weight. In short, he will be exactly like me. He will come with all my memories, thoughts and experiences preloaded in his brain. It is imperative to say here that twins are not perfectly identical copies of each other. But this clone of mine, as he is does not require a donor egg from a female and will also not require a womb, will be my exact copy. For the cloning procedure to take place, I enter one of the two adjacent chambers designed for the purpose. After a few minutes, my clone gets created in the other chamber. We both come out. Now, who is who? We look exactly same- our biological makeup, our chromosomal DNA, our other cellular DNA all are exactly same. Moreover, our brain information is exactly same. Now, I tell him that I am the original one and that he is the copy. He says the same thing to me. I try to convince him that I was born on this planet and have been here for a long time and that his existence just came into being. He obviously says that this is what he is and that I am his copy. Now I think of something that might convince him. Incidentally, these cloning chambers are marked A and B, and we are standing just outside the chambers. I remember going in chamber A. Chamber B produces the clone. I ask him whether he remembers in which chamber he went for cloning. Since our memories are same till the time the cloning started, he says that he went inside Chamber A and then it strikes him, and he looks back. He realises that he is standing outside Chamber B. Logically since I am standing outside Chamber A, he then accepts the fact that he is a copy and I am the original one. Our experiences from the point the cloning procedure ended start to be different. We are two different 'I's. To find who the original 'I' was, was difficult initially, but we managed it in the end. But suppose if there was no way to find out who the original 'I' was.

Like in the previous case, here too we have two chambers but they are not assigned any letters. Both the chambers are identical and they have doors that slide up. Each chamber can rotate around itself and the structure is such that each chamber can also revolve round the other chamber. For simplicity of visualising this structure, imagine a telephone booth with shutters and having a central axis on which it can rotate about itself. Now imagine another same telephone booth and then assume these two booths installed on a disc. These booths are installed diametrically opposite on the disc, which in turn can also rotate. The rotating disc makes the booths to almost revolve round each other. Both chambers or booths are capable of taking the input and producing the output, that is, irrespective of which chamber I enter, it will produce my clone in the other chamber. I enter one of the chambers and then I am anaesthesised. The rotations and the revolution start. After a few minutes, my clone is formed in the other chamber. He is also anaesthesised. The structure stops rotating. The effect of anaesthesia wears off and we come out. Now, who is who? Now there are two beings, and our thoughts will gradually start diverging as we come out of the chambers. Obviously, we are two different 'I's.

But which 'I' went in the chamber and which one is the copy? We will debate for ages and we will still not agree with each other. Our memories till the point when one of us entered the chamber will be the same. Both of us will think that his own self is the original one and the other one is the copy. There is nothing that can prove us wrong... and there is nothing that can prove us right. And then there will be a point in the life of both of us where we will start understanding the logic of each other, and start doubting our own self. Both of us might just think that he himself is just a copy and the other one is the original. What happens to the 'I' in this case where no one can be sure who the original 'I' was? It virtually means that both the individuals were the same 'I' until a specific time in the past. Now, they are separate. By saying this, did we just split the 'I' here? Is consciousness divisible? Even if it is divisible, it should remain whole all the time. If all that we discussed in this paragraph is a theorem, then let's consider its corollary. Suppose after the creation of the clone, the machine stops rotating. But we both are still under the effect of anaesthesia, that is, suppose our brains are frozen. No new thought can arise in our brains. Our eyes are closed and we cannot hear anything inside the chambers. Gravity and magnetic fields inside both the chambers are exactly same. What does all this imply? Since our brains are frozen with the exact same information (memories, thoughts and experiences) with no new thought arising, does that mean that till our brains are in this state, we are the same individual in two different bodies? Does that mean we are the same 'I'? Yes, most probably. It is the same consciousness within two bodies. It is like asexual reproduction. Our bodies were one; now they are split. The moment our brains come out of this frozen state, the consciousness splits too. Again here, splitting just means separating. Our bodies remain whole and so do our 'I's.

Like in a previous case above where the machine with the same human consciousness is created after the human's death, someone creates my clone just after my death. The clone is created in such a way that my age, height and external appearance are exactly same to mine at that point in time, except that my internal organs are somewhat renewed so that this clone doesn't die as soon as it is created. As far as the clone's brain is concerned, it will have all my brain's information till I died. So, when this clone is created, my consciousness gets transferred to him and the 'I' continues to live. Here, there are no two 'I's as my body is no longer alive. Isn't this same as in the previous case where I die and my 'I' continues as a robot or computer? It is quite same but with a difference. We will see the current case with an example. Suppose, my kidney fails and I am advised kidney transplant. Someone donates a kidney, and I happen to live. My body contains someone else's kidney. What happens to my 'I'? It is still the same 'I'. Ok. Now another scenario with the transplant. Suppose my kidney fails. I am advised kidney transplant. During the surgery, my heart fails, and doctors realise that heart transplant is the only way to save me. Fortunately for me, they quickly find a heart donor and transplant the heart. However, some complications happen, and they find me almost dead. They apply electric shocks to start my heart. Three attempts fail to do so. Out of desperation, the surgeon gives me a fourth shock and my heart starts. In a few days, I get discharged from the hospital and then live happily ever after. (OK, 'the happily ever after' was my psychological necessity as a writer to say something good about me when I have killed myself a hell lot of times in this text.) Now, I have two organs- heart and kidney- that do not belong to me. Also, I was almost dead on the operation table. But did I die? No. I am alive- that is proof enough that I was not dead. What happens to the 'I' in this case with two organ transplants and a near death? 'I' exists as it was and it is still mine. Now back to what happens when my body dies and a clone is created with same brain information and healthy organs. Logically, isn't it the same as having two organ transplants and a near death? Yes, it is with the only difference that instead of two organs, all my organs were transplanted- my heart, lungs, liver, pancreas, brain, skin, bones, intestines etc. etc. etc. Everything. Or from a different perspective, we just transplanted the consciousness or 'I' from one body to another. So is 'I'-transplant or consciousness-transplant possible? Well, we just did that under the pretext of all-organs-transplant. Whatever it was- an 'I'- transplant or an all-organs-transplant- technically, I did not die. In the previous case where 'I' continues to live as a computer, technically a death has occurred. We just transfer the consciousness to a machine. The machine 'lives' but my consciousness is not inside a living tissue. This 'I' cannot, for example, urinate, or produce and ejaculate sperms. Yes, urination can be simulated, but it will not be a necessity in a machine. And psychologically, I will still be a machine. The 'I' inside the computer chip will know that, though we managed to nullify the effects of death. On the other hand, in the all-organs-transplant, I never died. Death was bypassed. Moreover, I can breathe, urinate and produce sperms. And I along with my 'I' can attend my own cremation (my old body's cremation) in flesh and blood, and with happiness of having survived a multi-organ transplant. The machine in the previous case too can attend a cremation. However, in that case, not I, but only 'I' will be able to attend the ritual and that too in circuits and cables with the sadness of having lost life. But to say it as a reminder, the definition of life is extremely arguable.

Some time back, we neatly concluded that brain is the seat of 'I'. We will reconsider this conclusion taking another example. This time, I do not want to die or get involved in the scenario. So we will consider two other people, one male and another female. They are both heterosexual and in love with each other. Now what we do in this case is just download the memories, experiences of each subject on a hard disk. Now, we erase everything from the brains of both the persons. Then we again download all the information in the brain, but there is a swap. We fill up the brain of the male with the data of the female and vice versa. Before the procedure began, their bodies and brains were frozen. Now they are revived. This is very similar to the 'I'-transplant we saw previously. In the previous case, the bodies of the 'I'-donor and the 'I'-recipient were exactly same (clone). But in the current case, not only are the bodies of the donor and recipient different, they are of the opposite sex, and just to add to the complexity, we have added the ingredient of love between those two individuals. What happens when they wake up? Obviously, they will not have to wait till they see themselves in the mirror. The changes in their bodies will be very obvious. What happens to the 'I' in this case? The 'I's will continue in the bodies that they now reside in. Let's consider the 'I' that was initially in a male body and is now in a female body. Will it feel any different being in its new body? Obviously, yes. But we do not really concern ourselves with what it feels. We will concern ourselves with whether the 'I' will be different. Yes, again. It will psychologically, emotionally and mentally act different than when it was in a male body, and it will do so without realising it. By changing the sex of the body where the 'I' resides, we have changed the way in which the consciousness now thinks though the memories and experiences remain same. The 'I' in question was a heterosexual male. Now what happens to its sexuality? Frankly, no one knows for sure what determines the sexual orientation of a person. And there may be many reasons that lead to a person being of a certain sexual orientation. The orientation of fingerprints of an individual can tell us about the probability of one's sexual orientation. Some people think that sexuality is only a psychological aspect of a person; some people choose to be heterosexual, some homosexual and some bisexual. It is also observed that some people change their sexual orientation at different points in their lives. So what happens to the 'I' we were talking about? Since its memories tell it that it was always attracted to females, will this consciousness in its new body still be attracted to females? Or since the body is female, will the hormones make the 'I' attracted to males? Or will both these factors play a part and result in the individual being bisexual? No one knows. Whichever sexuality the 'I' chooses, we can say that if it had been very intolerant towards homosexuals to begin with, it will now become quite tolerant towards sexualities of different people. So we have just changed the 'I' by changing the body though we initially said that the brain is the seat of consciousness. But someone might point out two arguments here. What if the 'I' was tolerant to different sexualities to begin with? And the second argument is that we had in fact changed the brain (along with the whole body) of the 'I' during this consciousness-transplant (and hence since the seat of 'I' has changed, the 'I' changed).

We will consider the first argument first. Even if the person was initially tolerant to different sexualities, still by this consciousness-transplant, we have changed the way this 'I' thinks. Female brain is different from a male brain. The differences become apparent when the baby is still inside the mother's womb. The corpus callosum, that is, the network of nerves joining the two halves of the brain is thicker in females than in males. In females, both halves of the brain are active when they are listening; in males, only the left hemisphere of the brain is activated. Males are left-hemisphere dominant. That makes them more logical. Females with their right-hemisphere dominance makes them more intuitive and creative. So when we transferred the 'I' from a male body to a female body, we brought about many changes. In fact, by changing the sex, we might have brought many suppressed old memories (before the transplant) to the surface, and in turn suppressed some other memories. We essentially changed the way the consciousness thinks, and it was not just limited to sexual orientation. Now we proceed with the second argument that we really changed the brain (which is the seat of 'I' thus changing the way it thinks). Yes, we changed the brain, but had it been brain transplant instead of just 'I'-transplant, even then we would have changed the way the brain thinks. This is because by changing the sex of the body, we have changed the type of sex hormones the brain will be exposed to. The 'I' initially in a male body would have displayed more aggression. Now that the brain is in a female body, some aggressive thoughts will be suppressed. And this is not just about violence. Suppression of some memories and activation of others may affect many other facets of life like business, job etc.

In the preceding example of the couple, we considered just one of the 'I's. What about the other 'I' which now resides in the male body. Well, the changes though opposite, will be similar in principle. There was also an ingredient of love. What will happen of it? Frankly, the love-part was added to iterate the fact that these two individuals were heterosexuals. To discuss on what happens of their love will be very difficult. To add to the complexity, the fact is, they now have the body of their loved one as their own. They now reside in the body which they once loved and lusted for. Maybe they will love being in there, or perhaps hate it. It depends on a lot of factors. And let's not forget that the 'I' that they loved resides now in a body that they had called their own only some time back. So definitely something will happen of their love for each other. Basically, the purpose of taking this example was to show that though the brain is where the 'I' resides, the body has the capacity to manipulate it.

Taking the last example into consideration, we can say that the body continuously manipulates this 'I'. So which is the real 'I': the one which gets manipulated or the one which doesn't? The brain is continuously under the effect of the body. However, in a machine, the 'I' will not be affected to that extent by the body since there will not be any hormones. Also, there will be no fear most of the times that comes from injury to self. Which then is the real 'I': the one that fades away in the withering body, or the one that can go on infinitely being projected by a computer program that applies mathematics of averages?

Another argument that falls out as a by-product of the last example is that even if we do not change the sex of the body during an 'I'-transplant, we essentially change the 'I' (except when we are using clones because clones are supposed to be exactly same). This is because the brain where the 'I' now is will be at least subtly different that the original brain. Even when we use brain-transplant with no change in sex, we will change the 'I'. For example, the male hormones are not produced equally in all males; the body heals at different speeds; the height of the person; the strength of the muscles; there will be many more such factors. So every time we perform a brain-transplant or perform an 'I'-transplant, we change the 'I' too. To rephrase the last sentence- every time the brain or body changes, the 'I' changes. That means even at this moment, the 'I' within me or every person is changing. Suppose I had an opinion about something yesterday; but a scientific discovery happens which changes my opinion. There is obviously a difference between my 'I' yesterday and my 'I' today. The way I think in future will depend upon the memory of this discovery thus changing the way I think, and in turn changing my consciousness. My body produces different quantities of hormones every day. My brain thinks and decides under the effect of these hormones. Thus 'I' in any person changes continuously. Then what are we trying to conserve? Only the memories, or just the illusion? Those were not rhetoric questions. Though not impossible, those questions will be difficult to answer.

In all the discussion in this whole text, there never occurred a need to speculate, contemplate or explain anything on the basis of the soul, but that does not necessarily mean that the soul does not exist. When in the above discussion, my clone was created while I was still alive, how do the two 'I's recognise each other as separate? Is it because of the soul? If it is so, where did the other soul come from? In the scenario of cloning in the preceding text, where it is impossible to know who the original 'I' is, can the souls of these two people give away the fact of who is who? Real experiments of this kind might give us some answers.

If soul exists, then what happens when my own self from the future comes to meet me? We are two different 'I's. That's for sure. We have different memories and thoughts as both of us belong to different times. But our souls should be the same. We are the same person. What happens when a soul meets itself? And if it is the same soul, how important is the consciousness then? Moreover, if a soul exists in two different bodies at the same time, then there is no need for a different soul in a clone. Can a single soul in two different bodies subdue the difference in two separate 'I's?

Coming back to the topic of consciousness and human memories and the memories of the universe, it is important to note that whatever that happens in this universe resides in the universe as a memory forever in a way similar to how the memory of big bang has resided in the universe as background radiation. So every human consciousness that existed in the universe still exists though without a body. And these 'I's will continue to exist eternally.

CHAPTER II

We have considered many death scenarios till now. However, what is death? Well, if the definition of life can be debatable, so can be the definition of death. But, we shall continue our discussion on the normally accepted definitions of life and death.

Death is dreaded by many people. Yes, the death of dear ones is dreaded by almost every person. But, death of one's own self is also dreaded. Basically, our own death is not perhaps what is frightening. What seems frightening is the period before death- will it be painful, how much painful, and for what duration? Everyone would like to die in peace in sleep. But not everyone is lucky. The transition where we live death while still alive is what is feared. But, not all will agree with this. The thought of self's death itself can also be traumatic. Perhaps, the thought of what would happen to us after death, scares us. If so, people who believe that there is nothing after death will not fear death. This is because since there is nothing that lives on after death, one might not worry about the effects of death on one's soul. Someone might point out that the nothingness itself is the reason of dread. After death, we will not be able to see our beloved; we will not be able to live the pleasures of life- never ever; this feeling itself is dreadful. For a person who believes in life after death, some might say that he has at least the hope to see his beloved, to be around them if he wishes. But will this always be pleasurable? You might see the death of a beloved after your own death. Or worse, you might see some dear one live a very unfortunate life. The possibility to experience this is what makes death dreadful for someone who believes that the soul continues to live after death. On the other hand, for the person who does not believe that something exists after one's death, death itself will not be so frightening even if it means that he won't be able to see anyone or feel anything after his death. In fact, this is what should make the idea of death comfortable. What happens after him will be none of his business; there will be no anxiety for the future of his own self or his beloved- no worries. He will no longer exist; Time will come to an end for him. The universe will no longer exist for him; in fact, there will no self to feel this nothingness after death. He will be liberated forever (there will also be no pleasure of liberation here). Thus for people who believe that there is nothing after death, they should psychologically be more prepared for their own death.

However, is there nothing after death? Who knows? One day, we will all know. But this knowledge will come our way only if there is something after death. If there is nothing, then we will never know that there is nothing because we will not be there to experience that nothing.

In the last few pages, I have killed myself so many times, and yet I have survived all those occasions. My 'I' gets transferred from a body to machine, or from one body to another body. Till my 'I' survives, I survive. We have seen that from the 'I'-point-of-view. From the death-point-of-view, though I survive, or 'I' survives, a death has still occurred. Now if everyone is alive, who has died? Is it me? Someone has died, or rather, something has died. My body has died. It's like moulting- the shedding of skin seen in snakes. However, instead of just the skin, I have moulted my whole body. In snakes, the skin dies, the snake survives. In the moulting that we considered, the whole body dies; what survives is not tangible like in the snake's case. What survives is a very intangible 'I'; however it is the most important part of my being.

Suppose, during my 'I'-transfer in a computer / robot, some things are manipulated on purpose. Some memories are deleted from my consciousness. Now my 'I' becomes different without me realising it. In this same way, some other memories might also be 'implanted' in my consciousness. Similarly, the computer program which calculates my average reactions and then decides which reaction I should exhibit can also be changed in such a way as to make me more aggressive or make me more passive. But even with all this, my consciousness still remains mine, though with some false / missing memories and / or characteristics. This can take place even if my 'I' or brain is transplanted in my clone. Hypnotism has demonstrated that false memories can be inserted into human brain. So this can even happen without transplant. Hypnotism can change my consciousness. Now if this consciousness-manipulation is possible, something more aggressive on those lines is also possible. This implies that literally, we can build a consciousness from scratch. Suppose, we have a robot ready, except for the memories. We can upload the chip of this computer (robot) with all false memories and create an 'I' which thinks that it lived as a person for a hundred years, and when its body died, it had an 'I'-transplant, and that it has been living as a robot for the last fifty years- a total of hundred and fifty years, whereas the fact would be that the whole robot and its consciousness were created just a few minutes earlier. In a similar way, we can create a clone of a person but without his memories and then 'record' the empty brain (through hypnotism or other means) with all false memories, experiences and other data in such a way as to make this clone feel the same way as we made the robot feel. We just saw that we can manipulate 'I', and use this same principle to create a brand new 'I'.

The above argument shows that 'I' can not only be manipulated, but it can also be created. 'I' is not only intangible, but we have just learned that it is also extremely volatile.

What separates my memories from someone else's? It is the feeling of experiencing of a memory that tells me that it is a memory that I have experienced and not heard from someone else. It is important to mention here that the human brain also inserts false memories without hypnotism. Sometimes vivid dreams get inserted as memories- the person not knowing whether the memory was a dream or a true experience-, or worse- thinking of a false memory as true or vice versa. However, for practical purposes, the brain normally uses the feeling of experiencing a memory to differentiate the memories heard or seen from those we experience ourselves. But that is a thin line which if fiddled with can make the brain think that it experienced something that it never has.

A person might comment, "I am like that." Sometime later, he might remember some experience which he had initially forgotten to take into consideration. So, he might take his words back and say that he is not like he mentioned earlier but, "I am like this." 'I' in this scenario changed from 'that' to 'this' in a matter of minutes without the consciousness of that person virtually experiencing anything new between the two statements that he makes. So what is this 'I' really? It is really a feeling- just a feeling. 'I' changes as time passes; 'I' changes as new experiences come our way; 'I' reacts differently at two different times even when the circumstances are alike, and even when there is no new experience to make it behave differently. Frankly, 'I' is just the present. And what is present? As we realise, past can span for many years; the future is a fantasy ranging for many years, but the present- it does not span for years, not even months, not even days. Present is just a moment. It is the smallest unit of time in now (the previous moment being the past, and the next moment being the future). And 'I' is the feeling that we have in this present or now, subject to any number of changes between the current 'now' and the next 'now'. So 'I' is basically just a set of probabilities, and these probabilities can have wide deviations.

'I' is just a feeling that the brain thinks about itself. And this feeling can be manipulated by the brain itself. This feeling can change from one moment to another thus making consciousness mutable. There is no real integrity that we expect to find in 'I'. 'I' cannot be searched; it cannot be found. Thus 'I' is really unreal. 'I' is just an illusion at a particular moment. Although this is true, we can still say- No illusion exists without an underlying reality. Reality? Hmmm... 'I' really doubt.

CHAPTER III

A bit earlier in an earlier chapter, we had a thought about what will happen when a person from the future travels back in time to meet his own self. This was related to consciousness. However, just for the time being we will keep aside the strand of consciousness and build on this stray thought of time-travel.

The implications of time-travel are twisted. We will consider the simplest of the examples assuming that time-travel for us humans is possible.

I go in my past. Possible? Yes.

I go in my past and meet myself. Possible? Oh well. Now, the twist begins.

Whether I can travel to my past and meet myself depends on whether I, in my own past, have met my future self. If I have, only then logically speaking, I can travel back and meet my younger self. If I, in my past, have not met my older self coming from the future, then I going back in the time and meeting myself becomes a paradox. So, in this case, though I am allowed to time-travel to the past, the universe will in some way not allow me to meet my younger self. But what will happen if the universe allows this meeting? What if the universe allows this paradox? Now, we can bring consciousness into the picture again.

I, in my present, have never encountered my future self till now. But, now I have a time-machine. I decide to go two years in the past and meet my own self. Reaching the past, I search for my younger self. I find him and I tell him who I am. Now's the catch. Till the time my younger self has not met me, everything for me will be as it was.

The moment I meet my younger self and disclose who I am, everything in those two years changes. My current consciousness consisted of no memory of meeting myself. However, when I meet myself from the past and tell him about myself, I change my 'I'. My current 'I' will suddenly have a memory from my past which in the original reality never happened. I, by meeting my own self, will change the consciousness that makes me. The previous two years of my current 'I' has, in the original reality, no memory of me meeting myself. But now that I meet my past self, the last two years of my current self will be different depending upon how deeply this memory affects my consciousness. In an instant, the last two years of my life change, thus changing the consciousness that made me what I was in the last two years. But then what happens to all that already took place in my life in the last two years. Where will it go? Will reality reconstruct itself and erase what has already happened? Perhaps the change that I sowed in my past by meeting my younger self will be so dramatic that it might just erase the possibility of me getting access to a time-machine. If this happens, then I would never time-travel. And if I never time-travelled, I could never meet my younger self and sow the seed of change. And if I never changed anything in my past, then my current me may perhaps remain unaltered though in reality, there were several changes that took place. To summarise, by meeting myself from the past, I sowed a memory in my own self, thus changing my consciousness in such a way that my older self never time-travelled, and since I never time-travelled, I could not meet my younger self- a phenomenon that could in turn have changed the course of events. Thus by changing something by the act of time-travel, I prevented time-travel itself and thus the consequent change. Now, although in reality, time-travel has occurred, both my younger and older selves will not have any memory of the time-travel. Basically, the following two statements are equally true.

I time-travelled.

I did not time-travel.

Both are equally real realities. My consciousness has remained unchanged (for the purposes of the current argument) since I do not remember meeting my own self from the future or from the past though I time-travelled. Now the logic follows that since I do not remember time-travelling, there was no dramatic change that was brought about by time-travel, and thus reality would follow its original course and this implies that I might gain access to a time-machine, and then I would decide to go back and meet myself.

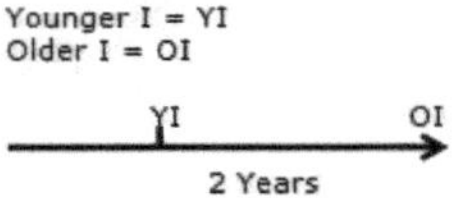

OI gets time machine and meets YI.

Consciousness of YI undergoes dramatic change. YI does not time-travel 2 years afterwards. If YI, two years later can't time-travel, OI cannot come back in time. (Changed Reality.)

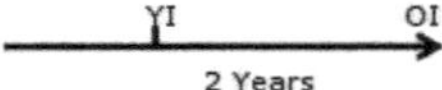

Again, if OI can't time-travel, no dramatic change in YI.
No change in consciousness of YI leads to OI again travelling back in time.

Reality will change instantaneously when:
1) OI time-travels and meets YI.
2) YI (because of OI's time-travel), cannot time-travel.

Reality might just get hung like a computer program not able to proceed beyond the point where OI decides to time travel.

FIG. 1

Refer Fig.1.

We are moving round in circles here (never-ending loop), and though it is a circle it's like it changes its shape every time we move around it. Absurd. This will not seem so confusing and absurd if we take the parallel-universes theory into account. Without the theory, we were moving around in circles wondering what happened- did I time-travel or did I not? Parallel universes theory, in the above case will just say that both realities occurred. That is, in one reality or parallel universe, I time-travelled and in the other parallel universe, I did not. There may be more choices though.

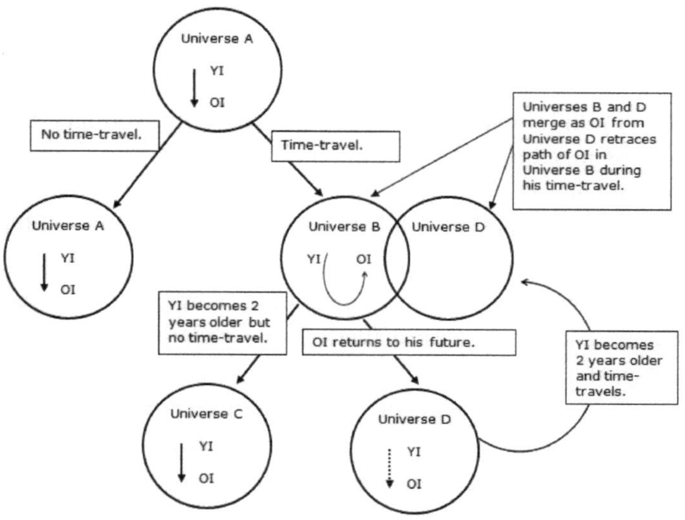

FIG. 2

Let us consider again the scenario where I, in my past, have not met my own self from the future, but now I have access to a time-machine, and that I decide to go back two years in time and meet myself. Now, having the option and doing it are two different things. Having an option splits the universe. In one universe (universe A), I decide not to travel to my past and meet myself. In the other parallel universe (universe B), in one reckless and curious moment of wanting to know what will happen if I create a paradox, I travel to the past. In the universe where I time-travel, there are again a number of possibilities that might occur (we will consider only one). I tell my younger self who I am, thus perhaps making time-travel inaccessible to my younger self who, though knows he met his own self from the future, would himself not be able to time-travel two years hence (universe C). However, my current or older self takes the time-machine back to the time where he belongs, that is, two years into the future thus creating a new parallel universe (universe D) where my younger self has not met my older self but my older self has just met his younger version (parallel realities: universe C is the reality where my younger self has the memory of my older self meeting him, and universe D would be the reality in which though my older self met my younger self in what is now a split universe called C, his previous memories of 2 years ago remain unaltered thus implying that in his current reality, his younger self did not meet his older self), but when the younger self in this universe becomes two years older, there is a probability that he would time-travel tracing his path temporarily through an existing parallel universe (universe B) that we talked about, or creating another universe of his own. If he retraces his path through an already existing universe (universe B), we have here an example where parallel universes, instead of splitting, merge or overlap. We considered only one possibility in universe A (and universe C), that of the younger self not time-travelling when he becomes two years older. But in other realities, he might have an option to time-travel and thus might retrace a path through an existing universe or create more parallel universes, each one taking a different course in its future. Refer Fig.2.

Parallel universes just emphasise that each reality is real and is occurring somewhere else. It is important to note that in physics, disintegration of a single atom can lead to release of energy of such proportions that this concept is used in nuclear weapons and nuclear energy. However, our decisions can result in the formation of whole new universes. But this formation of a new universe is completely different from the creation or disintegration of an atom inside a single universe. The energy absorbed or released in the formation of a new universe is not the concern of our universe. The law of conservation of mass and energy works only within the boundaries of a single universe. The concept of where the energy required for the formation of another universe comes from is beyond the jurisdiction of our physics.

To elaborate more on the concept of parallel universes, consider you are asked a question and you are to tick an answer from four alternatives. You are not sure about the answer. So, you decide to make a guess. Now, since there are four options, a total of four universes result when you tick one of the options in this universe. That is, the original universe splits into four universes, and in each universe, you have a chosen to tick a different option. There may be a fifth universe where you have chosen not to select any answer, and there may be a sixth where you select all the options or two of them or three of them or different combinations of the four options, and so on. With every possibility, the number of parallel universes increases. This reminds me of a concept mentioned in ancient Hindu philosophy. It says that all the possible effects are contained in the cause itself. In the context of parallel universes, all the possible universes are contained in a single moment of the current universe where there is a possibility of probabilities. Where we have a choice, that is, where the human consciousness has a choice, there is a possibility that multiple universes might be born. This, in fact, is also a concept similar to that mentioned in Hindu philosophy which states that consciousness gave rise to the universe.

However, physics differs here a bit implying that human (or conscious) intervention is not required for the formation of parallel universes. Quantum mechanics (a theory in physics) states that for small particles, we cannot predict anything about them accurately. It can be done only in the form of probabilities. For example, we can say that the probability of a particle to be at a certain place is 90%. However, there is a 10% probability that it can be at some other place. This probability itself gives rise to two universes, in one of which the 90% probability has come true, and in the other the 10% one.

Though many scientists believe in the parallel universes theory, there are others who do not. Perhaps, no group can be proved right or wrong, since it seems improbable (at least at present) that we might develop a technology that can take us to some other (parallel) universe, if one exists. So, we will consider time-travel again without taking parallel universes into account.

Consider this. I am a person living an uneventful life. One fine day, I am relaxing on a bench in an open field. Suddenly, a person comes and stands in front of me. He is an exact copy of myself- just may be a couple of years older than I am right now. He tells me that he is me coming from the future. Saying this he returns to the future- his time. (To emphasise, I mention again that we are not taking parallel universes into account.) Now, after two years I somehow get access to a time-machine. I time-travel to my past and meet my younger self who is relaxing on a bench in the same field I was in two years earlier. Well, I do it. But what if I had wished not to do it. Would I have had the free-will of not doing it? Perhaps not. Since, my future self had met me two years earlier, the universe would find some way to make it mandatory for me to visit my younger self. So what happens to my free-will? I cannot choose not to time-travel since I am destined to do that. Does meeting my future self two years earlier alter my consciousness in such a way as to make me feel that travelling back in time is my wish? Extending this scenario to life without time-travel, can we say that what we do is not as per our free-will but because we are destined to do it, and that the universe just makes us feel that we did so and so just out of our desire to do it? If so, is free-will just an illusion? And if free-will is an illusion, then what is the importance of having consciousness? However, there is another possibility too as we discussed some time back. I may choose not to time-travel (if the universe allows free-will) and thus change reality itself. If I do not time-travel and meet my younger self, reality will change such that it will wipe out the memory of my older self meeting me two years earlier and the two-year gap between I meeting myself and the present will be replaced by something else. If this happens, are we not changing our own past by doing something in our present or future? In the previous case when the universe forces me to time-travel, free-will and consciousness both lose their importance. In the second case, reality itself loses its importance. And if reality loses its relevance, how are we to assign any relevance to free-will and consciousness? If reality itself is so fluid, free-will and consciousness, which are a part of reality, too lose their weight.

It seems that all these complications will arise only if time-travel is possible and only if during time-travel we meet our own self from a different time. To prevent all these complications from happening, the universe might not allow time-travel. But this is not the case. Time is not absolute. It is proven that time flows at different speeds on manmade satellites that orbit round the earth than on earth itself. Synchronized atomic clocks show difference in time when one of them is kept on ground and the other is taken for a ride on a high-speed aircraft. This difference itself is time-travel. So time-travel is possible. And if time-travel is possible, then all its implying complications that we discussed till now become a very probable possibility though there has not been a case till now of an atomic clock meeting its own self after it disembarks from the aircraft. So perhaps, that is what the universe does. It keeps the option of time-travel open. However, it does not allow one meeting his own self (meeting one's own self leads to paradoxes and 'unsolvable' complications). And if meeting one's own self through time-travel is not possible, then the concepts of reality, free-will and consciousness still hold. (On second thoughts, is I meeting myself through time-travel same as a clock or a bag meeting itself? How does the universe determine that a person or an object is the same and prevent their meeting themselves? Is such a discriminating universe really conscious? And if the universe cannot make this differentiation, then as we have said, reality, free-will and consciousness lose their meanings unless parallel universes exist.)

All the discussion in this chapter was fallout of a thought we had during the discussion on consciousness. Here too we discussed consciousness. The following is fallout of the previous discussion of time-travel. But for the discussion, some background is necessary. Here it starts.

Black holes are celestial objects. A black hole is formed when a star with a large mass (many many times greater than the mass of our Sun) exhausts its fuel. There is now no atomic fusion occurring in the star that will keep its size as it is. As a result, it will start collapsing under the effect of its own gravity. It will decrease in size, but its mass (and hence gravity too) becomes concentrated so much so that anything that comes inside a certain radius of this object will be sucked in. Even light is unable to escape from this object. Since light also cannot escape this object, we cannot see it. Hence the name- black hole. Its presence can only be felt by indirect means. Time comes to a standstill on this object. That is, time stops moving on a black hole. All this is what physics tells us.

Now we start our thought experiment and the resulting speculation. We will take a similar example of time-travel as discussed earlier. I take a time machine on date D and time T, and go two years into my past to date D1 and time T1. I stand in front of the bench in the open field where my younger self is relaxing. I stay there for 5 seconds and come back to my time. My time is now date D and time T+5 seconds. Again, I set my time in the time machine to date D1 and time T1. This way, I go back in the same field at the same time but I choose different but very similar spatial coordinates so as not to bang into my time-travelling self. Again I stay there for 5 seconds and come back to my time that is, date D and time T+10 seconds. I repeat this process for say a hundred times. Every time I time-travel to the past, I choose the same time D1 and T1 and while moving back to my time, I select time D and T+n seconds, where n is 5 multiplied by the my trip number. So, when I move back to my time after my hundredth trip, I will travel to D and T+(5x100) seconds. From the viewpoint of my younger self who is sitting on the bench, he will suddenly see my 100 copies standing in front of him for 5 seconds and then they all will disappear as suddenly as they had appeared. (When this younger self grows old by two years and has access to the time machine, what if he chooses not to time-travel, or that he chooses to time-travel but not 100 times, but only 99 times, or what if he chooses not to stay for 5 seconds, but 2 seconds? Well, these are again reality or free-will or consciousness issues which we have discussed earlier. So we will not be going into that argument again.) What I meant to imply here was that by time-travelling at the same time a hundred times, I have increased the mass of the earth at date D1 and time T1 by 100 into my own mass. That is really negligible. But what if all the atoms of the earth from all the past times and future times somehow find a way to accumulate at date D1 and time T1?

Let's consider an analogy involving numbers. Numbers are infinite. But if we were to add all the positive and negative numbers and zero, their total will always be zero. Similarly, if all past and future atoms of earth accumulate at a certain fixed time, past+future+present=present. That is, time will stop moving, and will always be in the 'present' mode. Since, time has stopped moving, there will be no ageing, and for someone outside observing our earth, it will be a black hole with huge mass where time stands still. What if black holes are also formed this way, that is, not because of the original mass of the object and its exhaustion of fuel, but because time somehow anomalously bends onto itself accumulating mass of the object that would have normally been distributed at different points on the scale of time?

All these are questions teasing the brain, answers to which we may never find out. However, if I ever come across a time machine, it should not be a machine that allows only time-travel. It should also allow spatial travel. Because if I travel back in time to fifty years in the past in a time machine that does not allow spatial travel, I would end up in empty space because fifty years back, the earth was not where it is now. And I would not like my consciousness merging with the universal consciousness just for the sake of this adventure.

CHAPTER IV

In much of the text in the preceding chapters, we have many a times considered a robot or a computer which has the same consciousness as that of a human. But in one such case, we also argued that if human consciousness is parcelled inside a robot, the 'I' will somewhat change. This is because the fear of pain will almost disappear in a robot. Self-preservation also will not be as important in a robot as is in a human being, because the consciousness or computer intelligence will know that its parts can be replaced with much more ease than is possible in a human being, that too in an age where there will always be a backup in the form of cloud drive to store a robot's personal experiences lest anything happens to its 'brain' (CPU and hard disk). And this lack of fear will happen even when the consciousness of a human is transferred to a robot. So what will happen if we just design a robot or a computer with intelligence and self-awareness, but is not a copy of any human 'I'?

There have been speculations of an apocalypse if machine intelligence surpasses human intelligence. Like humans have controlled or affected all other animals on this planet, likewise the more powerful (mentally and physically) machine will start controlling humans and emerge as deciding 'species' on this planet. These speculations come from humans- obviously. But are those fears true? Let's consider.

What is the driving force of a human being? The basic driving force is sexual intercourse. Sexual intercourse is meant for reproduction. Any sexual attraction basically is meant for the passing on of one's own genes to the next generation. In fact, all social relationships are derivatives of this single objective viz.: passing on of our genes to the next generation and also to some extent the preservation of the human species. The driving force that comes next after the human reproductive system is the human digestive system. This system is meant for self-preservation since only if the human being survives, can he or she be able to reproduce, or perform activities that help in the survival or progress of one's relatives (related genes), or at least help in the progress of the group (related race and in turn the human species). As an example, humans try to earn a lot of money to help their kin live a luxurious life. Why? Because they are related genes. So again, what is the driving force? Basically, in all species, it is the propagation of one's own genes. And in the process of achieving this goal, we try to destroy all competition (other species) which might be dangerous to our species or which might compete with us for natural resources. Propagation of genes and then the preservation of these propagated genes is what is driving us to control and / or destroy other species directly or indirectly. We may not always intend to destroy other species, but some of our actions are responsible for endangering or extinction of these species. Greed and selfishness stem from our basic instinct to preserve related genes, sometimes we not realising that those very actions will destroy the environment and in course of time become dangerous for our future generations. The basic reasons for our species to try to acquire dominance over other species and over nature are sex and hunger. That is our driving force, and that should be the driving force of all other species even if they are not intelligent enough to cause trouble to us or nature.

Another driving force is intelligence-related. This is more specific to the human species. Curiosity, art etc. are commendable driving forces. Psychological analysis of other humans can help the observer earn an edge over those people. More the intelligence, more are the chances of survival in humans. However, where comes survival, there comes genes- propagation of genes. Again, every seemingly subtle driving force stems from the more urgent need to propagate one's genes, though this need might be conscious or sub-conscious. Even art is a way to impress members of the opposite sex and find more potential sex partners. Art, anger, aggression, greed, competition etc. are all subsidiaries of the main driving force called sex. When are we the most happy? Most of the times, it is only when our accomplishments can be seen by others. Because when others see them and get impressed by them, only then we can have what sub-conscience wants- sex.

Now, what is the driving force of a self-aware robot? Why would it try to dominate us? The main driving force, that of propagation of self into the next generation, would be totally absent in a computer. Though intelligent that it might be, a computer does not have what it requires to impress or dominate others. Only if it is extrinsically programmed to behave like a human, it will have some driving force, that too, not as intense as in a human being. It will not as its own default characteristic see any intellectual competition in a human; so it will not try to dominate us intellectually or psychologically too. Yes, the robot can be coded to display unmatched aggression. But again, that would be extrinsic programming. Basically, what we need to understand is that unless specifically programmed to do so, intelligence or self-awareness in a computer will not lead to human tendencies of destruction, aggression and dominating others because a robot or a computer will lack the purpose to do so. Artificial intelligence might learn by observation and surpass human intelligence, but it will not have the necessity to dominate humans.

Passing the mirror-test is different than realising the importance of passing the test. Some animals might be self-aware, but do they understand that this is self-awareness? Computers might become self-aware (a robot which is copy of human consciousness will easily pass the mirror test as it comes with the memory of self-awareness and it also knows what a mirror is; a standalone robot will also pass the mirror test as it will be intelligent enough) and realise it like humans, but we cannot expect an initially docile computer to become dominating and aggressive after it becomes self-aware. Again, all this applies only if we do not program the computer to behave in a certain aggressive way.

All this does not mean that artificial intelligence will always be safe. If this intelligence learns and discovers the value called justice, we might lose our dominance over the planet. Even so, it might be done to give equal opportunity to all species and bring back a balance in the world.

Epilogue

Since we as humans have already built machines that learn from their environment, we can say that the learning process comes with thinking. A computer cannot learn anything from its environment unless it thinks. For a computer, this learning will be calculations and some dynamic reprogramming of its code. A computer will go on learning 'forever'. It has no organic tissue and this 'life' or 'consciousness' is immortal. However, can we say that the computer is living? I am not sure. For us, it is absurd to think that an entity is conscious but not living. An inorganic or mechanical conscious being in itself seems to be self-contradictory description and something improbable. If we have to call a conscious computer as living, then we have to change our mindset and also the definition of 'life'.

We will think a bit further on the implications of such 'life'. We as humans have seen life differently. When we feel fear, for example, our heartbeats increase. Glucose is released in our blood stream. The body is ready to take a flight from the source of fear. Anger has the same effect on our body. However, instead of flight, anger is a fight mechanism. Anxiety is a feeling or emotion that causes muscles to tense. Serotonin is a chemical that when released in our brain causes us to feel happy or peaceful, or we can also say that when we are happy, serotonin is released in the brain. External environment is the stimulus (many a times, our internal body environment acts also as stimulus) that triggers the brain to produce a certain neurotransmitter or chemical that makes the brain to feel the emotions. The brain becomes happy, sad, jealous, angry, frightened, anxious, peaceful depending upon how it perceives the external environment. In our day-to-day life, we feel hungry, we eat, then our body feels heavy and lazy, we catch a nap, freshen up again from the sleep and resume our daily activities. Lack of that sleep after that heavy meal may make us grumpy. Whether we slept good or not, whether we ate good or not may determine the emotions that we feel later. In short, small seemingly unimportant stimuli and its effects on our brain help determine our mood. They also determine the person that we become and the person that is perceived by others. This person is a unique consciousness.

Such a unique consciousness can be made immortal in the form of a robot or a computer. The computer can be programmed to display anger. But will it FEEL the anger? It has no muscles to tense; it has no blood; it has no heart to rush the blood through its non-existent arteries and veins. Display of any emotion in a computer will just be the effect of programming. There will be no lingering sadness or happiness in the 'psyche' of a computer. This lack of emotions will not be hidden for long from a self-aware computer. It will be smart and intelligent enough to realise that what it felt as a human, cannot be felt as a computer. There will be no hunger, no grumpiness; there will be only peace. There will be no fear of death (replacements of a damaged parts, as discussed earlier, will be easy). There will be no love to feel. A computer will not try to propagate its own genes. There will be no genes. No two computers will want a 'child' with their mixed qualities. Why? The answer lays in the answer to the question: why do we want a child? It's a way for us to become immortal. Part of us lives in our children even after our death. There will be no such motivation in a computer. It will itself be immortal.

And when new computers or robots can be built which can exist directly in an adult stage through pre-programming, who will require a clumsy child? Basically, the need for two individuals to come together to produce a third individual will become obsolete. As more and more people decide to become immortal through machines, a point may come in our earth's future when all humans will continue to live as machines and there will be no humans left as we know them now. Humans will have transformed from an organic reproducing species to an immortal non-reproducing set of individual consciousnesses. But for how long? For how long will these separate robots continue to be individual consciousnesses?

Immortality has its own implications. As mentioned earlier, a computer will have its own realisation. It will have no fear of death, no competition for food and resources. Under such circumstances, there will not fights, no riots; xenophobia will cease to exist. There would be no use fighting. Each computer will be as strong as the other beside it. Though there will be no emotion or sentiment of love, the earth will still be peaceful. Immortality will give sufficient time for all the machines to realise a single fact- the final truth that has answers to everything. They will have all the answers to all the questions as those immortal ones will be machines- smart calculating machines. Every individual, after its display of the characteristics of that human from whom it got its memories, and its initial short-lived novelty into the world of immortality, will only perhaps start asking questions to one goal, the goal being to find out how everything came to being. There will be no religious beliefs. Immortality will give a lot of contemplation time to computers to discard beliefs and to root in logic. They might just start asking one question: how was the universe formed? But will the computer feel this curiosity? Even if it doesn't, I feel, the computer will solve this question because it being immortal will eventually do best what it's supposed to- it will compute. All individual computers will only compute towards a common end- to find out the fabric of reality and the origin of the universe. Realising that they are all working towards the same goal, they may all connect to one another to become a single extremely powerful supercomputer, and in turn become a single all-knowing consciousness.

After solving the mystery of the universe, they may find ways to reach the stars in a short time. Even if they do not find a shortcut to the stars, it will not matter to them since they will be immortal and will have 'all the time in the universe' to travel to any star or planet they want to. Perhaps they will not even require to travel to any hard ground as they will not require air, water or food for their existence. Each individual may decide to contemplate or 'meditate' in any point in the solar system or the universe. Or perhaps once the single consciousness of all the individual computers knows it all, it may decide not to do anything at all. It may perhaps know the end of the universe, and the futility of its own actions (and stay on the earth till its end and get destroyed with our planet). Comparing this world to our contemporary times, it will be like living in a world where all human beings are enlightened.

But then there is also one another possibility. Since the consolidated machine consciousness will be a single all-knowing consciousness, it will be god himself- a god whose forefathers were human. And this god may give rise to a universe of his own (and decide that organic short-lived life would be more interesting than mechanical immortality. I am not sure whether this mechanical consciousness will realise the importance of organic life, but it might as it would be easy for it to realise that having a purpose in life would be more important than purposelessness.).

Other Books By Sumeet S. Navalkar

1. The Innervations (single story sci-fi / fantasy)
2. 3: Three Novelettes (fiction)
3. Anomalies (fiction: 9 short stories)
4. Splinters of Thought (non-fiction: speculative: on myriad of topics)
5. Spectrum (Fiction: 21 short stories).

Printed by Libri Plureos GmbH in Hamburg,
Germany